ISBN 978-0-282-34022-3
PIBN 10848260

This book is a reproduction of an important historical work. Forgotten Books uses
state-of-the-art technology to digitally reconstruct the work, preserving the original format
whilst repairing imperfections present in the aged copy. In rare cases, an imperfection in
the original, such as a blemish or missing page, may be replicated in our edition. We do,
however, repair the vast majority of imperfections successfully; any imperfections that
remain are intentionally left to preserve the state of such historical works.

English
Français
Deutsche
Italiano
Español
Português

www.forgottenbooks.com

Mythology Photography **Fiction**
Fishing Christianity **Art** Cooking
Essays Buddhism Freemasonry
Medicine **Biology** Music **Ancient
Egypt** Evolution Carpentry Physics
Dance Geology **Mathematics** Fitness
Shakespeare **Folklore** Yoga Marketing
Confidence Immortality Biographies
Poetry **Psychology** Witchcraft
Electronics Chemistry History **Law**
Accounting **Philosophy** Anthropology
Alchemy Drama Quantum Mechanics
Atheism Sexual Health **Ancient History**
Entrepreneurship Languages Sport
Paleontology Needlework Islam
Metaphysics Investment Archaeology
Parenting Statistics Criminology
Motivational

SIMPLE DECORATIVE LATHE WORK

EXAMPLES OF DECORATIVE TURNING EXECUTED BY THE AUTHOR.
oto by]   FIG 63 (*See p. 85*).   [*A. G. Rider, Winchester*.

SIMPLE DECORATIVE LATHE WORK

A PRACTICAL HANDBOOK ON THE
CONSTRUCTION AND USE OF
THE ORDINARY TURNING
LATHE FOR THE PUR-
POSE OF THE
ABOVE ART

By JAMES LUKIN, B.A.

AUTHOR OF "POSSIBILITIES OF SMALL LATHES"
"TURNING FOR BEGINNERS," ETC., ETC.

SEVENTY-THREE ILLUSTRATIONS

LONDON
GUILBERT PITMAN, 85, FLEET ST., E.C.
1905

E.S.C.

INTRODUCTION.

THE object of the writer in penning this unpretentious little volume is to supply those who desire to practise the simpler operations of Ornamental Turning with a useful handbook to that fascinating work. The apparatus needed is inexpensive, and much of it may be home-made by any one who is a fair hand at metal work—such light metal work in this case as may be fearlessly done upon a lathe with traversing mandrel, which is now obtainable at a very low figure.

For those whose means allow them to aspire to the higher branches of Ornamental Turning there is the splendidly illustrated work of Mr. Evans, now being re-issued by the publisher of this little volume, and which, I have no hesitation in saying, is the cheapest and most complete work ever published upon this interesting subject.

Advanced age and illness must be the present writer's excuse for the many faults and imperfections of this his last attempt to help the learner over the initial difficulties of Ornamental Lathe Work.

J. LUKIN.

Romsey, December, 1904.

LIST OF ILLUSTRATIONS.

CONTENTS.

CHAPTER IV.

THE UNIVERSAL CUTTER FRAME.

CHAPTER V.

THE OVERHEAD.

CHAPTER VI.

THE DIVISION-PLATE AND INDEX.

CHAPTER VII.

THE ECCENTRIC CHUCK.

CHAPTER I.

THE ORNAMENTAL SLIDE-REST.

THAT ornamental turning is a fascinating pursuit, and capable of endless elaboration, is a fact that no enthusiastic lathe-man will dispute. It is not altogether legitimate *turning*, but is rather a system of mechanical carving, effected by various ingenious contrivances fitted to, and driven by, the ordinary turning lathe. The apparatus required for high-class work of this kind is very costly, and is, for the most part, not well suited for a lathe under 5-in. centre, and although some of it can be fitted to a lathe of 3-in. centre, it is necessarily in such case of a somewhat diminutive character, and the division plate on the mandrel pulley is less legible than a larger one, nor will it, of course, allow of an equal number of holes. All the fittings are, therefore, of slighter construction and are less capable of standing the wear and tear of constant use. The 5-in. centre lathe admits of more solidity in its several parts, and the necessary chucks and apparatus can be used with greater freedom and confidence. Nevertheless, to those to whom money is an object, I may say, without hesitation, that a great deal of very beautiful work can be accomplished on a lathe

B

of only 3-in. centre, even though it may not be fitted with traversing mandrel.   Taking in order the various essentials of an ornamental turning lathe, we shall have to speak of

The Division Plate and Index,

The Slide-Rest,

The Eccentric Chuck,

Eccentric Cutter,

Drilling Spindle,

Vertical, Horizontal, and

Universal Cutter Frame,

with the small tools necessary for the above.

Of other more costly and complicated apparatus we need not treat at present, and no doubt to some of my readers even this will appear a sufficiently formidable list.

The Slide-Rest: This piece of apparatus is altogether different from that needed for metal turning. The latter may be, indeed, pressed into service, but unquestionably there are serious drawbacks connected with its use.   It is, in the first place, heavy and clumsy (the main screws are not always very accurately cut), the tool-holder is ill-suited to receive the shanks of the several cutter frames, and (which is a still greater fault) it cannot easily be fitted with certain guides and stops which are absolutely needed in order to produce satisfactory work   This piece of apparatus, if it is of an up-to-date kind, is a costly affair, but in its simplest form need not be such an expensive item as to prohibit its purchase by amateurs of moderate means.   It is not, indeed, a very difficult piece of work for home manufacture, and a fairly skilful mechanic may easily construct one for his own

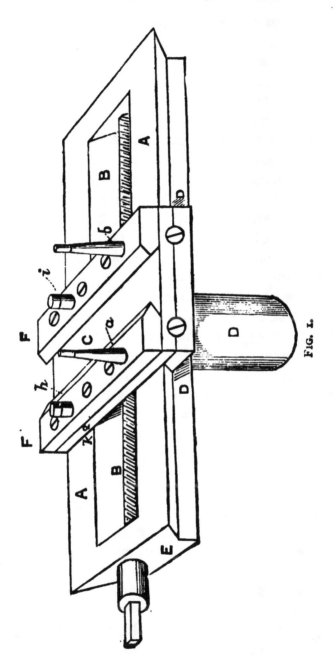

FIG. I.

use. I may very probably return to this matter of home construction, but at present my subject is rather "how to use" than how to make. The slide-rest for ornamental turning consists essentially of a rectangular frame of cast iron, 12 ins. to 14 ins. long by about 2¼ ins. to 3 ins. wide, with a turned stem below the centre to fit into the socket of the sole plate, which is similar to that of a hand rest—the frame taking the place of the ordinary tee,—a leading screw accurately cut, of, generally, ten threads to the inch, traverses this frame lengthwise, and carries a winch-handle at one end. Upon this main frame is fitted a slide with nut underneath it, through which the main screw passes, and by which it can be traversed in either direction. This slide carries the tool-holder, which is in this case a rectangular open box or trough, in which the shank of the tool lies, and in which it is fixed by a pair of small clamps. I give a drawing of this simple rest in Fig. 1. The price complete of such an one for a 5-in. lathe would not be more than £5 or £6, and rather less for a smaller-sized one. In Fig 1. AA represents the main frame through which passes the leading screw BB. The slide C, with its nut underneath, is fitted with two bars bevelled to fit the similarly shaped edges of side on which it slides under the impulse of the screw. One of these steel bars is seen at D. head of the leading screw is graduated into ten divisions, reading against a line engraved upon the surface of the frame. One complete turn of the screw will move the slide with its tool-holder $\frac{1}{10}$ of an in., and one of the divisions will correspond to a movement of $\frac{1}{100}$ of an in. On the plate C of the slide are two double-bevelled bars F F, one or both of which is slightly adjustable, and between these slides the tool-box represented in Fig. 2. This is a solid casting of brass or gun-metal, with two lugs (G G) at one end, bored and tapped to receive a pair of screws, of which one is shown in place, and the other is represented in Fig. 3, and will be further explained presently. The channel E E, in which the shank of the tools lies, is

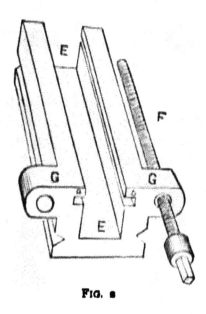

Fig. 2

carefully planed out to $\frac{9}{16}$ square, this having become a standard size for the shanks of the tools and apparatus used in ornamental turning. In the original rests there was no standard thread used, but the head of the screw was arbitrarily divided according to the rate or pitch of screw selected. The result was the same, but did not admit of certain refinements now considered indispensable, viz., the decimal system applied to the width of the tools as well as to the leading screws of the various apparatus. At *ab* will be seen two narrow channels running the entire length of the tool-box. These are for the reception of a pair

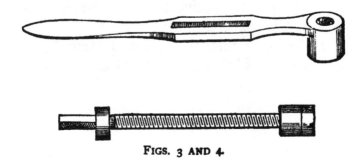

FIGS. 3 AND 4.

of clamps, of which one is represented in Fig. 5. These consist of a conical pillar of steel, with a square plate at the bottom, which is filed up accurately to fit the grooves or channels just spoken of. These

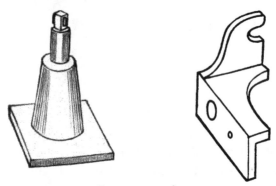

FIGS. 5 AND 6.

pillars are turned cylindrical for about $\frac{1}{2}$ an in. at the top, and are then bored and tapped with a fine thread to receive a screw which has a small washer attached below its head, being squared to take a small spanner.

This screw causes the washer to press upon and secure the shank of the tool. One of these clamps is short, but the other is extended, so that when in use its top is even with that of two similar pillars (*a* and *b*, Fig. 1) which are screwed into the guide bars of the tool-slide. A lever (Fig. 4) containing a hole and slot fits upon one of these fixed pillars and upon the top of the longest tool-clamp, and by this means the tool-box can be moved to and fro in the slide C, carrying with it the fixed tool or other apparatus. As before stated, there are *two* screws passing through the lugs of the tool-box (Fig. 2), one of which regulates the depth of cut by coming into contact with a stop (*i*, Fig. 1); while the other, when in use instead of the lever, comes against a similar stop (*h*). This stop has a flat filed upon it with a mark as a reading line engraved on its top to read against the head of the screw shown separately at Fig. 3. When the lever is used this screw is not in use, but if it is required to give a slow and steady motion to the tool-box, a small bridle, which is cut out of a thin plate of steel, is attached by a screw and steady-pin at *k*, Fig. 1. This bridle (Fig. 6) spans the screw just behind its cylindrical head, so that the latter lies between this bridle and the fixed stop. Thus, when the screw is made to revolve by its winch-handle, the tool-box is caused to advance or retreat, but it cannot advance beyond the limit required to give the tool its intended penetration, because the stop screw comes into contact with its own stop. There are other ways of effecting and controlling this movement of the tool-box, but that which is here described is generally adopted, and is quite as simple and as easy to arrange as any of the more primitive and now obsolete methods. The bridle (Fig. 6) is an exceedingly neat contrivance, and does not add to the complication of this modern slide-rest.

Although not actually belonging to the original design, the elevating ring and fluting stops may be considered almost essential, as well as the cradle, to be presently described. At the same time, really good ornamental work was undoubtedly done before these were invented, and the ornamental rest, as already described, is very far superior, even in its most simple

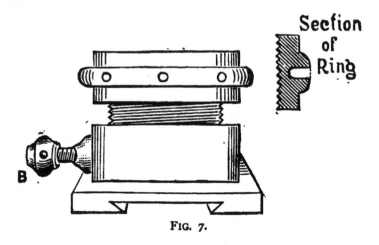

FIG. 7.

form, to the ordinary rest for metal turning. The elevating ring is fitted to the top of the pedestal, which has a screw-thread cut upon it. It is a ring of brass or gun-metal, about an inch in height, with an inside screw to fit that on the pedestal. When screwed down to its lowest position its upper edge is level with that of the pedestal, but when unscrewed it has the effect of making the pedestal somewhat higher, and as it is in contact with the under-side of the main frame, it will raise this more or less so as to bring the point of a cutting tool exactly level with the lathe centres. This is in many cases absolutely essential. In old times the stem of the frame was raised by hand in its socket, and then clamped by the side screw B, like the ordinary tee used for plain turning. It was difficult to do this with perfect accuracy, and when it became necessary to alter the position of the rest, it was almost certain to drop below centres again before being secured by its clamping screw. With the elevating ring once raised to the proper height, the clamping screw may be loosened without any possibility of the stem of the rest dropping down further into its pedestal. This ring is generally made with a projecting

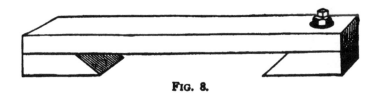

FIG. 8.

moulding, milled to give a hold to the finger and thumb, but holes are also drilled round it to take a lever or " Tommy," as it is generally called, as it should fit fairly tightly so as to prevent it from being moved accidentally. It is represented in Fig. 7, where a cross section of it is also shown. Another almost necessary addition to a rest is a device to regulate the traverse of the tool-slide along the main frame, and is more especially used in cutting flutes. These generally need to be carried to an accurately defined distance, and nothing looks worse than one or two of a set longer or shorter than its fellows. Fluting stops (Fig. 8) are made in pairs to fix on the main frame each side of the tool-slide at any required distance. For absolutely perfect accuracy, a small screw is tapped into each so that its point shall come in contact with the tool-box. This is a refinement seldom needed in actual practice, but it is so easy to make and use that it is hardly worth while to omit it. Of course, however, each such addition adds to the cost of the ornamental slide-rest.

CHAPTER II.

THE ORNAMENTAL SLIDE-REST (Continued).
THE DRILLING INSTRUMENT.

BEFORE parting with the slide-rest and its fittings, I may as well point out a method of dispensing with the fluting-stops as a matter of economy, although the proposed substitutes are confessedly a makeshift. I had myself a rest without these orthodox additions, and contrived to do some very fair work with the following arrangement: I sawed out two strips of box-wood and planed them truly rectangular, of a size to fit nicely but stiffly between the cheeks of the main frame, filing or rasping out a groove along one side where it would come in contact with the main screw. All I then had to do was to saw off two pieces of just such a length as would permit the tool-slide to traverse the necessary distance in either direction. Thus the length of every individual flute was absolutely determined by the contact of the tool-box with the end of the box-wood stop. Of course the latter was made just long enough to abut inside the ends of the main frame, and could not be pushed further. I have used this simple device to determine the traverse of the tool in a metal-turning rest, and I can therefore recommend it as a cheap and efficient substitute for the brass fluting-stops. The latter, in their best form, have pointed screws tapped into them at the centre, the extreme points of which are the actual stops which come in contact with the tool-slide, and allow of even more perfect adjustment than could be gained by allowing the tool-slide to touch the actual brass-work of the stop. This, if such extreme refinement is desired, is easily added, and adds nothing to the cost ; but it is a question whether it is a really necessary addition to the apparatus. The turner will soon find that even with a stop, however well fitted, something is left to delicacy of touch when executing really fine work. It is one thing for the tool-slide just to come into contact with the stop, and another to press hard against it, and in two adjoining flutes, in the finest work, the difference would be quite noticeable to the eye of an adept.

The drilling instrument is perhaps, from the simplicity of its parts and the facility with which it can be applied, the first to be placed in the hands of a learner for the purpose of decorative turning.   It has taken different forms ; but that which is now universally adopted by the best makers is of the following construction : A bar of steel from three to five inches long, planed accurately rectangular to $\frac{9}{16}$ in. square, is drilled through lengthwise, and the corners of the bar are just eased off in the lathe at each end, or turned down at the forward end to give it a more finished appearance.   Each end is then turned out for a little distance, and hard steel collars are fitted and coned out to receive a steel spindle, also hardened where it is in contact with the collars.   Sometimes one end only is coned ; but this is a detail for the manufacturer rather than the turner.   The absolutely requisite condition is that the spindle shall run truly and with perfect freedom, and shall remain true after a good

FIG. 9.

dcal of use.   Some method of compensation for wear and tear is therefore always provided in well-made instruments of this kind. At one end this internal spindle is bored to receive the shanks of the small drills or cutters used with it. At the other end is a pulley of one or two speeds to take a cord from the overhead pulley which drives it. It is essential to the proper action of this tool that every individual drill be turned to shape *while in its place in the drill-spindle,* a tedious and rather difficult job which some makers of cheap tools very often shirk ; but the work done by it in such case is certain to be inferior to that which can be effected by a properly-made drill.   To make this clearer I have drawn the tool complete in Fig. 9, with some of the drills in process of formation (A, B, and C), and I do so because it is advisable for the sake of economy that a turner should make such tools for himself.   The finished drills are shown in Figs. 10, 11, and 12.   Now the central point of Fig. 10 must revolve in a truly axial line with the centre of the spindle, or it will cut a small circle.   Likewise, in Fig. 11, which is a fluting drill, the same necessity exists, as it does in Fig. 12 and nearly all others.   The drills are therefore cut off a suitable bar of steel wire, fitted in the socket of the spindle, and the tool is then held in the slide-rest and driven by a cord from the overhead, the hand-rest being fixed close to the steel blank. Thus the drill-stock becomes a small lathe in which to form its own tools by hand-turning. I may observe that the steel blanks will need careful annealing; and when finished the little tools must be hardened and finished on small conical

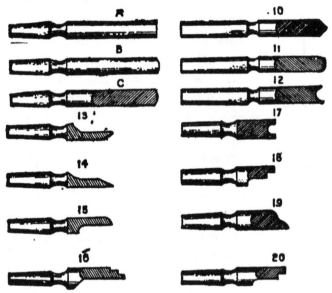

FIGS. 10 TO 20.

and other laps till the edges are keenly sharpened and *polished.* This polish is essential to best work. Each blank is first turned cylindrical, and a centre-mark is left on the end. This centre is the guide by which to file the drill to its ultimate shape. It may ˙ pointed, rounded at the extremity, filed into a ꞁi-circular or other hollow, or to any other form ꞁh by its revolution when in use will give the

desired form of flute moulding or beading; but this
could not be done unless the exact centre of revolution was first secured. When speaking of fluted
designs I shall have to allude to this more in detail.
In the sketches A, B, and C, the drills are represented
proportionately longer than the finished tools. Kept
short, they are less liable to spring when in use, and
they are easier to turn. A sharp graver is the best
tool to use in their formation, and they are then to be
filed to shape and subsequently finished on brass
cones fed with oilstone powder if they are hollow like
Fig. 12. An inspection of Figs. 13, 14, 15, and 16,
also of 18, 20, and 21, which is drawn to a larger
scale to make this principle clearer, will show that in
all of these one-half of the cutting end has been filed
away and the part which will be in action reduced to
the axial line or centre of revolution. These often
cut cleaner and leave a better finish on the work than
the drills whose edges are not thus reduced. Fig. 18
would cut exactly the same kind of bead as Fig. 17;
but as only half the width of edge would be in action,
the resistance would be less and the drill would work
more easily. Instead of filing away the drill to the
diametrical line, a very common plan is to bend the
metal when hot, like Fig. 21, which has the same
effect, while admitting of greater strength and breadth
of edge. This is known as a stepped drill, and would
cut a raised bead like a circular flight of steps. Such
a drill cannot, like Figs. 10 and 11, be traversed along
the work while cutting, as it would destroy its own
pattern during such traverse; but such a drill as Fig.
22—a double quarter hollow—may be traversed, and
will leave a raised moulding of such section, as Fig. 23.

This drill, again, might evidently be also cut away on one side as far as the axis of revolution without altering the pattern produced, as seen in Fig. 24.

I am reminded at this point that, as I am not writing for experts, but possibly for those who have never seen this kind of work done on the lathe, it may be as well to explain that the term drilling applied to ornamental turning does not necessarily imply that the work is pierced by a drill, but that these tools are used to cut the material to a certain depth and to remove a portion of it partly by penetration and partly by longitudinal traverse, the result

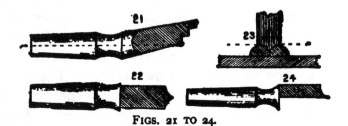

FIGS. 21 TO 24.

being in general to leave upon the work a reverse or fac-simile of the form of the cutting edge of the drill. A drill, for instance, with a flat chisel edge may be made to cut a rectangular channel partly by penetration and partly by traversing the tool along the work by means of the slide-rest.

For the present, however, my purpose is to describe the tool as commonly constructed by the best makers, together with a few ordinary forms of drill. I may here, therefore, add that sometimes the drilling instrument is double-geared to add to its power when used, as it often is, for such work as grooving metal taps or drilling other work requiring more power than can be obtained without such addition.   This, however, need not be illustrated here, as our subject is not metal-work but decorative turning.   There is, however, one other pattern of drill-stock which is sometimes more convenient than the form illustrated, though it is in other respects identical with it.   In this the pulley is attached in front just behind the socket which receives the drills instead of at the further end.   In using this the cord from the overhead which drives the tool is at a less angle, and therefore less liable to slip off the pulley when the work is of large size, and is being ornamented round the circumference.   The cost of the tool is the same, and it is perhaps on the whole the better form, although in practice there is probably less advantage than at first sight is suggested.   The forms given to the cutting edges of ornamental drills are not very numerous, as may be conjectured from the nature of the operation.   Nevertheless, what a lathe-maker would term a full set would comprise several dozens; and seen, brand new, in their cases, they look as it were a pity to sully their appearance by putting them to practical use. The reason they are made in such numbers is that there are several sizes of each pattern, the difference between any two being expressed in decimals of an inch.   This accords with all screws of ornamental rests and other apparatus, so that one complete turn of the handle of the slide-rest will move the tool a distance exactly equal to its own width.   In practice this is extremely convenient, but it is in such particulars that ornamental work *at its best* needs so much money to carry it out.   In the present day a turner demands that he shall work with ease and

c

certainty, and with as little trouble as possible, **and** consequently he is wont to purchase such numbers of tools that the probability is he will never actually use a tenth part of them. At the same time, he has the joy of possession, and it is a pleasure to be able to show admiring friends all the beauties of his costly apparatus. But I may use an old quotation and say, " In the beginning it was not so." Yet we have illustration of very creditable specimens in old books on turnery of drilled and other work.

CHAPTER III.

THE ECCENTRIC, VERTICAL, AND HORIZONTAL CUTTERS.

FOR "scratched patterns," or geometrical designs incised upon the surface of hard-wood blocks, the drilling spindle may be used to a limited extent with crank-shaped drills. These (Figs. 27, 28, and 29) will, when revolved, describe circles of fixed diameters, the intersection of which in various combinations constitutes the patterns in question. If these are to be printed it is necessary to trace them very lightly on prepared blocks of boxwood, which may be obtained in various sizes from those who deal in engravers' materials. If cut deeply the printed lines and spaces are too coarse, and any beauty of design in the original is entirely lost. If, on the other hand, the block itself is to be kept as a specimen of incised carving the cuts should be deep, and the tools themselves keen and highly polished. The cranked drills are, however, not very well suited to these deeply-cut patterns, in which the ever-shifting lights and shadows constitute the chief beauty. For these the only efficient revolving tool is that to be now described, which is called the eccentric cutter.

This instrument (represented in Fig. 26) consists, like the drilling spindle, of a rectangular bar of steel bored through to receive a steel spindle, which is fitted to revolve in hardened collars. At one end is attached a driving pulley, generally made with two speeds ; and at the other is a small steel frame at right angles to the spindle, usually forged of the same piece of steel.   This carries a small tool-holder actuated by a screw of ten threads to the inch.   This being of small diameter is always cut as a double thread.   An inspection of the drawing will show that the upper surface of this frame is arranged to stand level with the axis or centre of revolution of the spindle, so that a pointed tool lying flat upon it and clamped by the tool-holder will, if set in rotation, make a dot, and not a circle, if brought to the centre

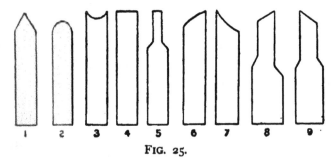

Fig. 25.

longitudinally by turning the screw by the milled head at the end.   The screwed hole forming the nut is partly filed away so as to allow the tool-holder sufficient vertical movement to allow it to clamp the tool securely.   Such a pointed tool shifted away from the centre of revolution by the little leading screw will cut circles of various sizes when put in revolution by a cord from the overhead ; and if such circles are caused to interlace a geometrical pattern will result. This cutting incised circles is, however, but one of the many uses of this ingenious and serviceable instrument.   Fig. 25 (1 to 9) shows a few forms of cutters used with this instrument, and these may be greatly multiplied by adding tools of similar outline at the cutting edge but of varying widths. It is by no means wise, however, to lay in a heavy stock of any such cutters or drills until facility in their use has been gained by practice. They are expensive to

FIG. 26.

buy, and not easy to make satisfactorily, and in practice not a great many are actually used. With a single dozen a good deal of handsome work can be accomplished. The same tools will serve for use with the eccentric cutter, vertical and horizontal cutters, and with the universal cutter, which is an embodiment of the two last-named in one instrument.

The vertical cutter, of which Figs. 30 and 31 represent two forms, has, like the previous ones, a $\frac{9}{16}$ shank to lie in the tool-receptacle of the slide-rest. At one end is either a forked or a plain part at right angles to the shank. This, as shown in Fig. 30, is bored, and has hard steel collars inserted, as in the drilling instrument, to receive the spindle carrying in a transverse mortice the cutting tool, which is similar to those of the eccentric cutter, and is made to suit either of these instruments or the horizontal or universal cutters, to be presently described. Fig. 30 will cut into narrow corners where the frame of Fig. 31 would touch the work, otherwise the latter is somewhat preferable, as it runs lighter—the spindle being mounted between centre-screws—no collars being needed.

The action of this lathe appliance is evidently to scoop out hollow channels by the rapid rotation of the little cutters of various sectional forms, those most commonly used having flat, chisel-like ends, or rounded ones, the latter also cutting flutes similar to those made by round-edged drills if the instrument is made to traverse horizontally as well as to revolve. It will not cut a flat as the eccentric cutter will, but the surface formed will more and more approach this form as cutters of greater length are used, giving to each cut a greater radius. At the same time, it must be observed that the further a cutter projects from the spindle the more likely it is to stick fast in the wood or ivory, and the more gently it must be advanced. If this form of instrument (Fig. 31) is used it is evident that the pulley will prevent a very short cutter from being used ; whereas in the form of Fig. 30 the pulley can in most cases be kept entirely out of the way. The section of a cylinder of which the side was thus ornamented would have the appearance of Fig. 32 ; but it is evident that the longitudinal flutes may be extended the entire length, or made to occupy only short portions of it, and that these may be made to alternate with others above or below, or cut so as to form spirals or other devices, the division-plate on the mandrel pulley determining their posi-

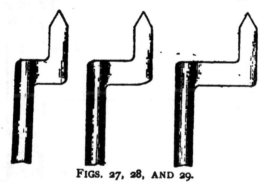

FIGS. 27, 28, AND 29.

tion and number. In the use of these revolving cutters driven by the overhead we have the secret of most of the specimens included under the common name of decorative or ornamental work. It is not, strictly speaking, ornamental turning, but decorating surfaces already formed in the lathe. In metal work a similar operation is denominated milling, which includes cutting the teeth of wheels and pinions, grooving, slotting, and many other mechanical operations needing the use of revolving cutters. The lathe proper simply holds the work in position, allowing it to be shifted after each cut, but the mandrel no longer revolves under the impulse of the fly-wheel. Simple as this appears, now we are accustomed to the arrangement, it was not brought into use for many a long year after the lathe itself had become a well-known machine. The earlier plan was to use a fixed tool and—by means of a special chuck on which the work was mounted—to give the latter a variable movement during its revolution. This is still effected by the eccentric or geometric chuck,

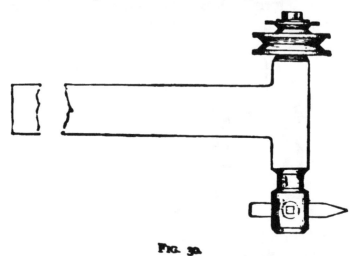

FIG. 30.

will be hereafter explained; but these are very ... ve, and on the whole less easy to manipulate ... the simple revolving cutters now being de .... As might be supposed, more complicated ... result from a combination of apparatus of kind—the article being mounted on an eccentric ... while a revolving tool is made to act upon ... surface—but this more complicated arrangement ... comes within scope of the present chapter. It ... at any rate, be delayed till a later page.

The horizontal cutter is no more than a vertical cutter-bar laid upon its side with the necessary addition of a pair of guide pulleys, or fair leaders, as they are sometimes called, over which the cord from the overhead is passed to the actual driving pulley. The spindle carries the same little cutters, and the only difference is that they work horizontally instead of vertically. The spindle is, however, also frequently fitted with small circular saws, or circular cutters, so formed as to cut wheel-teeth to the exact form necessary for use in clock or watch work or other machinery requiring the mathematically correct outline to enable them to work smoothly in rolling contact with their pinions. These do not need to be driven nearly so fast as the simple cutters already described, and the spindles of the drilling-frame vertical and horizontal cutters are often double geared to increase the power while reducing the speed. This arrangement, however, is not necessary for ornamental turning, although even for this it may be useful when very deep cutting is desired. All these additions to the lathe have undergone many changes of form, and may be modified to suit the requirements of the turner; but experience has led to the adoption of certain standard arrangements of their working parts, which have proved generally efficient, and these instruments are to be found at the shops of all lathe-makers who make a speciality of ornamental apparatus. In every class of machinery certain standard sizes have long been recognised as almost a matter of necessity, so that ornamental apparatus purchased at any one of the numerous manufacturers' establishments will fit the slide-rests and lathes purchased elsewhere;

but this is not the case with *chucks*, whether for plain or other work. These cannot be made to fit several lathes independently so as to run absolutely true. Each must be finished on the particular mandrel on which it is to be used. One other point must be noted in this connection, namely, that such a piece of apparatus as a drilling-frame is usually made with a shank $\frac{9}{16}$ of an inch square, because the tool receptacle of a slide-rest is now generally of this standard size, and the axis of rotation of the drill is arranged

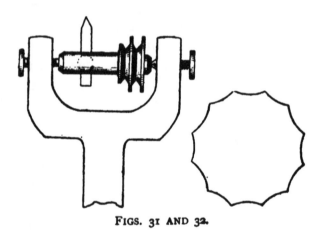

FIGS. 31 AND 32.

accordingly to fall exactly upon the line of centres of the mandrel and back-poppit. But this will depend upon the level of the tool-plate of the slide-rest, which may differ in lathes by different makers, especially if the lathe is purchased at one shop and the rest at another. Most slide-rests of modern construction of 5-in. centre will take a drill-stock with $\frac{9}{16}$-in. hank, and many lathes of 4-in. or 4½-in. centre will ike the same tool, *but it is not a matter of certainty*, the tool-plate, or bottom of the tool receptacle, may

E

be a little higher or lower than usual. It is always necessary, therefore, in the case of a drilling-frame, and only slightly of less importance in the case of vertical and horizontal cutters, to ascertain how far below the axis of the mandrel the level of the tool-plate is, and double this measure will be the size of the square of the shank of the drill-spindle. This will, of course, bring the point of the drill exactly on the line of centres. You can, of course, easily pack up the shank of a drilling-spindle with a parallel plate of metal, but there is no means of lowering it unless the slide-rest is made with an elevating screw. It is therefore absolutely necessary in buying such apparatus independently of the lathe to test its capability of accurate adjustment, or the tool may be found after purchase perfectly useless for its intended purpose. In my initial remarks upon this central adjustment I used the expression " of only slightly less importance in the vertical and horizontal cutters." This refers to the fact that the latter instrument admits of adjustment independently of the slide-rest on which it is used. There are (as in the vertical cutter) two centre screws of hard steel, which support between them the spindle on which the cutters are fixed. These are made sufficiently long to admit of a fair degree of traverse in the arms of the main frame. Thus the cutter may be raised or lowered $\frac{1}{4}$ of an inch or more to suit the height of lathe centres; and it will be found to make a great difference in the pattern whether the revolving tool is placed on, above, or below the line of centres. All these cutter-frames are exceedingly simple tools, but of extensive use, as will be more and more fully recognised as the turner advances in knowledge of his art ; and if he knows anything of pattern-making and metal-work generally, and has access to a foundry, he will find but little difficulty in making his own vertical and horizontal cutter, because there are no steel collars or deep boring, such as appertain to the drilling-spindle.

CHAPTER IV.

THE UNIVERSAL CUTTER FRAME.

THIS is a combination in one instrument of the vertical and horizontal cutters, but with the further advantage that the revolving tools can be set to cut at *any* intermediate angle. The stem is made, as before, $\frac{9}{16}$ in. square, unless, for reasons already stated, it is required of less or greater dimensions. This stem is bored lengthwise to receive the steel spindle, which is firmly attached to the brass frame which carries the spindle of the various cutters; and guide-pulleys, which can be wholly removed, direct the driving-cords when necessary. These guide-pulleys run quite freely upon their axle, as will be more easily seen in Fig. 34, so that they adjust themselves when in action, which will be the case when the instrument is used as a horizontal cutter or at an intermediate position. The main spindle is clamped by the nut B, and there is sometimes a graduated circular part at F. An inspection of this instrument will naturally suggest that it will so entirely take the place of the horizontal and vertical cutter that there can be little need to purchase these; but such is hardly the case, as will be readily discovered when practical work is in hand. The simpler and smaller instruments will always be preferable in the more delicate operations of ornamental turning; and it will often happen that the framework of the universal cutter will prove an obstruction by coming into contact with some projecting part of the work in hand. On the other hand, the facility with which the revolving tool can be made to cut at any desired angle gives the universal cutter a great advantage over the other two instruments. In reference to this I may mention that a clever mechanic now

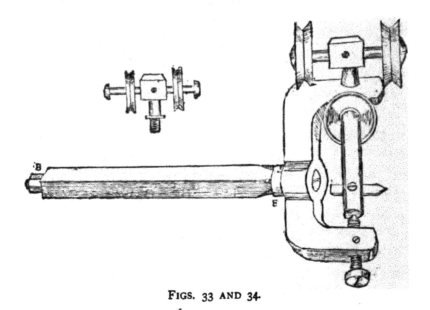

FIGS. 33 AND 34.

dead—one of those ingenious men whose active brains are for ever devising new expedients—suggested the introduction of round instead of square shanks to all this class of tools, so that the angle at which the cutters are presented to the work might be varied at pleasure. He also considered that the less skilful amateur would find it easier to turn a cylindrical stem than to file up a rectangular one.

He was himself essentially a latheman, and wished to make the lathe a universal shaping tool. I am speaking of the late Mr. Gilray, with whom I had a voluminous correspondence, and whose elaborate but very heavy and complicated lathe fell into my hands. The suggested innovation, promising as it was in some respects, entailed too great a departure from the accepted standard to be generally adopted. Not many amateurs make their own lathe apparatus, and with professionals articles of square section present no special difficulty—and to make cylindrical shanks to the several cutter-bars would in no way improve or cheapen these lathe appliances. The tool-holder of the slide-rest must also have been altered to secure these round-shanked tools.

Fig. 35, copied from an old French work, is another form of universal cutter, in which the rotation of the tool is effected by toothed gearing instead of a cord. I had one similar in my own possession some years ago, and it was perfect in action, but I objected to the humming noise inseparable from gears running at a high speed, and, rather to my subsequent regret, sold the instrument. This little tool was in use a hundred years ago ; so were the orthodox fluting stops to the ornamental slide-rest, the balance-weight overhead, and many other lathe appliances now in everyday use. The parts C, D, of this universal cutter are those which, like others of that date, fitted with V-grooves into the slide-rest. There was at that time no rectangular tool-box to receive cutters or tools with squared shanks, but only two bevelled guide-bars, between which fitted the main frame of each individual instrument. This particular tool had, as will be seen, a round stem fitting a hole in the main frame so as to permit the geared apparatus to be set at any angle and secured by a screw (H). Guide-pulleys are not needed, as the alteration of angle has no effect upon the cord from the overhead, which passed round the driving pulley (F) at the back of the main spur wheel. This is, in fact, a bevel or mitre wheel, driving the pinion on the axle of the spindle which carries the cutting tool. This wheel, shown again in Fig. 36, was dished consider-ably to allow room for the revolving cutter. There

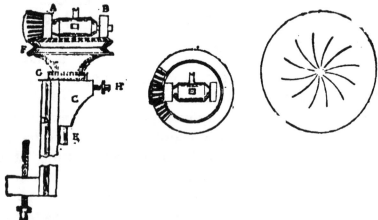

FIGS. 35, 36, AND 37.

is something wrong apparently in the drawing of the frame in both Figs., and A and B could not have been placed as shown within the gear-wheel, but I have copied it with fair accuracy from the original drawing. What I chiefly remember of my own instru-ment was its beautiful workmanship. The gears shown in the drawing are comparatively coarse, whereas those alluded to were fine, and most beautifully cut, running with the perfect smoothness of watchwork. Suppose, now, the stem (E) of this instrument to be lengthened so as to fit into the square stem of Fig. 33: the tool is at once modernized to fit the receptacle of the orthodox ornamental slide-rest, and we have a universal cutter with the advantage of no guide-pulleys. With regard to the frame, I think the parts A, B, supporting the revolving spindle, represent the square *tops* of the pedestals, the frame itself being at the back, and concealed from view by the other parts of the apparatus. The work done by these revolving cutters has the drawback of being exceedingly tedious to execute, although practice tends to diminish this somewhat. After each cut the lathe must be stopped, the index peg carefully shifted, and during the process the division plate needs the most careful scrutiny. Nothing can be more vexatious than a miscount, and it is unfortunately by no means difficult to make one. The result is necessarily an irremediable defect in the work, which may have taken many hours, or even days, to bring to the proper condition for the ornamental cutting. It is here that the 5-in. or 6-in. lathe has so great an advantage over a 3-in. or 4-in. one. In the latter the division plate is so small that it is very trying to the eyes, and there is very little choice in the number of holes, of which 98 is the highest easily obtainable. I have, however, submitted to Mr. Milnes, of Bradford, a device by which a large division plate may be attached to any lathe with traversing mandrel, and as he considers it a good and feasible arrangement I will give an idea of it, as I think it can be understood

D

without a drawing. The division plate, which may be 6 ins. to 9 ins. in diameter, is to be attached to a tube of brass or steel, which is to slip on in place of the long sleeve, which prevents the traverse of the mandrel when the screw-guides are not in use. It will be secured by the nut which ordinarily keeps the sleeve in place, so that no alteration of the lathe itself is needed. It can be fixed at either end of its own tube, or, if preferred, the sleeve itself can be turned down at the outer end to pass through the central hole of the division plate, which can then be fixed by a nut and small key. If it were not for the pin upon which the half-nuts are centered this large division plate could be placed close to the back of the mandrel headstock; but if this pin remains it must be fixed at the outer end of the sleeve. In either case it will be clear of the lathe-bed and is not restricted as to size, and it will admit plenty of holes, which can be marked at intervals by readable figures. These last might, I think, be filled with red sealing-wax varnish, to render them still more visible. It now remains to arrange an index. For this purpose we may take away the block of half-nuts at the back of the headstock, leaving the plain stud or pin upon which it is fitted. Let a short, stout ferrule or sleeve of brass be fitted at one end with the spring which carries the point of the index, and secure it with the screw which ordinarily secures the block of half-nuts. Such an arrangement may be constructed without the smallest difficulty, and the comfort of a large, easily-read division plate will amply repay the small cost of its manufacture. In connection with this arrangement— which, I may add, Mr. Milnes is prepared to carry out upon any of his 3-in. lathes with traversing mandrels —I would suggest a thicker division plate, capable of being drilled round the edge with a series of equidistant holes to receive two adjustable pins to form what is called a segment engine. Two pins fitted in the holes selected at any given distance apart come in contact with fixed stops, so that the mandrel (turned by hand) can only revolve a certain distance. The drill, or other revolving tool, can thus only cut out a groove to a predetermined arc of a circle, as seen in Fig. 37. The simplest way of arranging this that I ever came across was by attaching a piece of gut to a peg in the lathe-bed or headstock, with a loop at the other end slipped over one of the pegs of the division plate. Of course, such plan is a makeshift, but in this case it was the makeshift of a very clever and successful ornamental turner, and I speak of it because I am writing upon elementary work of this class in which inferior expedients must not be despised where the perfected apparatus is absent. The orthodox segment engine of perfect construction is an expensive affair, and is only fitted, as a rule, to lathes of the highest class. Apart from the dividing and drilling which lathe-makers will do at about 5s. per 100 holes, the construction of such a plate as I have described would present little difficulty to any amateur fairly skilful at metal-turning ; but it would, of course, need a lathe of higher centres than that upon which it is to be fitted. The cost of a brass casting is very small when it is of simple form not needing cores, and in this case the pattern is simply a circular piece of half-inch board, which any turner would prepare in a few minutes. Although, in fact, simple decorative lathe work is a special class of work, and the lathe for such work is of a light character compared to one intended mainly for metal, the amateur workman should not altogether eschew the practice of the mechanical branch of his art. He should be able, at any rate, to repair or even construct the simpler apparatus required, and thus save himself very considerable expense. Such work will greatly add to his skill in the use of purely ornamental apparatus.

CHAPTER V.

THE OVERHEAD.

I MENTIONED in the last chapter that some few, at least, of the supposed modern lathe fittings are described and illustrated in an old French work—the second edition of which dates back to 1816; the first was, I think, issued in 1800. The last plate in this second edition is a drawing of what is there called, by way of distinction, an English lathe, and is the only one which illustrates the overhead and revolving cutters, together with the simpler form of the slide-rest for ornamental turning.

This overhead, which I have copied, is to all intents and purposes the same as now made upon the balance-weight principle, in which only one cord is used. This arrangement is shown in Fig. 38. There is the single upright pedestal rising from the floor, or attached to the left standard—the bar overhanging the lathe-bed pivoted near one end, the pair of double guide pulleys, and the balance weight. The only difference between this and one of modern construction is, that a flat swinging bar is now generally used instead of a round one; but one is, in point of fact, as good as the other. In a modified form of this overhead, recently adopted for my own use, the swinging bar is hinged to the upright at the extremity, so that there is no short overhanging part, and a cord is attached to the other end, passing over a pulley above and descending to the weight. As the action of the weight is thus direct, it may be much lighter than in the former case. The cord may rise from any one of several holes in the bar according to the strain desired, in addition to which the weight itself is a piece of lead pipe slipped over the cord and supported by a short bit of wood, which, if placed parallel to the pipe, will easily fall through it, and when put into position at right angles to the cord will secure the weight in position. Thus one or more such weights may be used, and, if cased in a piece of brass tube—such as may be bought of any

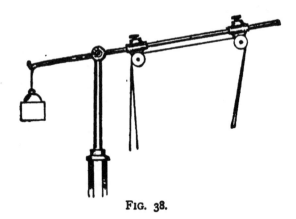

FIG. 38.

desired size at the ironmonger's—there is nothing unsightly about it, as the lead pipe is out of sight (see Fig. 39). The advantage or disadvantage—for it is more than anything a matter of choice—of this swinging-bar overhead, over that to be presently described, is that one cord only is needed (which passes from the fly-wheel to the drill or revolving cutter), whereas the other form needs two, in addition to a tightening apparatus, and the cost of this is also greater. It has, however, a neater appearance, and is capable of a more extensive application.

But so far as simple ornamental turning is concerned, in which the sole object of the overhead is to drive revolving cutters, the balanced bar is fully effective. Fig. 40 is a drawing of the usual overhead in which two uprights carry a shaft centered between them. On this is keyed a pulley to receive a cord from the fly-wheel, and it also carries a roller as shown, or a second pulley capable of adjustment, to take a second cord which goes to the pulley of the

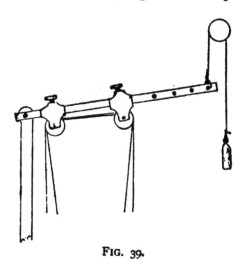

FIG. 39.

drill or cutter fixed in the slide-rest. The uprights are here shown as made of wood carrying the centre screws; but they are generally of iron curved over at the top. It matters little how the shaft is supported so that it runs freely, but the uprights must be placed forward enough to bring the shaft and roller over the pulley of the drill, as this will generally be nearer to the operator than the lathe-bed.

As it rarely happens, in ornamental turning, that the whole length of the lathe-bed is brought into use, there is no real necessity to extend the shaft of the overhead to that distance—a short iron frame suspended from a single support, or fixed to a beam above, will suffice for all ordinary work, and this may easily be fitted up by the amateur with the help of the local blacksmith.

Another consideration will determine, in some degree, the form of this useful addition to the lathe, and that is, the nature and position of the workshop. If the latter is in a room, which, perhaps, cannot be wholly given up to mechanical work, and if the overhead is solely needed for driving ornamental cutters, the neater and more compact it is the better. In this case the standards will be attached to the frame of the lathe, and either the balanced or roller overhead will probably be used. When, on the other hand, an independent workshop is available, with beams above, a stronger shaft, supported on hangers, may be preferably adopted, which may serve for such heavier work as milling, drilling, and other similar operations on metal, as well as driving ornamental cutters.

Then, again, it may happen that the amateur may wish to avoid expense by fitting up an overhead without professional assistance. In such case he may press into his service for standards, or even shafting, that very useful and easily-obtainable material, gas tubing, which will answer the purpose admirably, and will make as neat a framework as can be desired. The tubes can be obtained of various sizes and cut to any length, and can be united by straight sockets or knee-pieces, which can be bought ready-made, with screws, as can also be bought flanges with sockets, by which the frame can be screwed to the floor. If a length of such tube is to be used for the revolving shaft, it is only necessary to plug each end with steel. The frame of my own overhead is made in this way of iron gas-tube 1⅜ ins. outside diameter, and it is as firm and rigid as need be desired. Moreover, if work has to be carried on by artificial light, the frame can be utilised to convey gas—the burners being screwed into the uprights—single or double-hinged brackets being used as most convenient. With the latter the light can be concentrated upon the index-plate, or upon the work, or upon both at once.

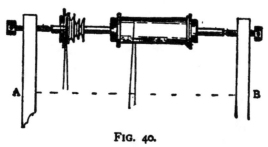

FIG. 40.

These iron tubes, which are not cast, but drawn upon a mandrel, will carry an excellent screw-thread, and short lengths may even be converted into useful chucks.

An inspection of a gasfitter's stores will, in short, bring to light many an article in iron and brass capable of conversion by the amateur turner to unexpected uses.

Whatever form of overhead may be preferred, some tension apparatus is needed to keep the cord just sufficiently strained to drive the selected cutter. In the balanced overhead (Fig. 38) this is fully provided for by the weighted lever and general arrangement of the apparatus; but in the case of an independent shaft,

there are two cords instead of one ; and although
that which goes to the fly-wheel is of invariable
length, and is once for all of determined tension, such
is not the case with the cord which passes from the
overhead roller to the revolving cutter.

The tension of the latter needs therefore to be
adjusted by some special apparatus. I here give a
sketch in profile of the two arrangements commonly

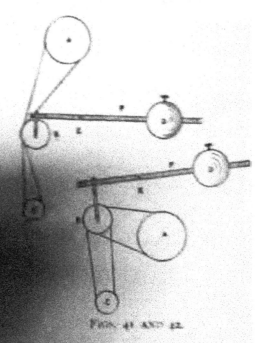

FIGS. 41 AND 42.

Fig. 41 (E) is the point of suspension or
of a lever (F), usually of flat iron, and upon
a weight (D), which can be fixed at any
the clamping screw tapped into it. The
of the bar carries a pair of pulleys running
a short axle fixed at right angles to the
roller or adjustable pulley seen in Fig. 40

is shown at A in Fig. 41, and the bar upon which the lever moves is situated at AB (Fig. 40), so that this apparatus is *below* the overhead shaft. The pulleys at B (Figs. 41 and 42) therefore press against and bend the cord by which the pulley (C) of a revolving cutter is driven, and the tension is regulated by the position of the weight (D) upon the lever. This is a very common arrangement, and answers sufficiently well in practice. Fig. 42, nevertheless, in which the apparatus is in all respects similar, is considered an improvement, and is now very generally adopted by the best makers. Here, it will be seen, the entire apparatus is placed above instead of below the main driving shaft and its roller (A). The cord passes up from the roller over the pulleys on the lever, and hangs down in a loop, and is thus always ready to be passed over the small driving pulley of the drill or other revolving cutter. The sketch is merely in outline, to make the arrangement of the tension apparatus as clear as possible. As usually made, the bar on which the carriage slides, which has the lever pivoted to it, is triangular, or is of square section placed anglewise, and is secured to the uprights by nuts at each end, the bar being turned cylindrical at its extremities, and cut with a screw-thread where it passes through the standards. The entire framework is thus very stiff, and vibration inappreciable, undue vibration being detrimental to best work. Every care is taken to make the revolving shaft run as lightly and freely as possible. The screws which form centre-points are of steel, hardened at their extremities ; the ends of the shaft are also of hard steel, and the centres deeply drilled to retain the lubricating oil. The roller is hollow, and is made of well dried and seasoned mahogany, and care is taken to make the shaft of sufficient size to stand the strain without bending, while keeping it as light as possible. It has even been proposed to let it run in ball-bearings instead of upon centre-screws ; but this is, I think, of questionable advantage.

I know but one other design of overhead, in which the roller and its axle are mounted in a square iron frame, but without any balanced weight or fitting previously described. The frame itself is suspended by springs from an overhanging support, which forms the upper goose-necked end of a single standard. In this, of course, the tension depends on the sustaining springs, which are adjustable by means of milled-headed screws like those of the safety valve of an engine. This is a neat arrangement, but the roller shaft does not exceed in length about half the length of the lathe-bed, and it is difficult to tighten the cord which drives the cutters without affecting that from the fly-wheel of the lathe, or pulling the frame out of the horizontal. Apart from this, no overhead can be neater, and many use it and approve it.

In this matter of overhead, therefore, there is ample choice of pattern, and the only real consideration lies between the single cord or the roller which needs an additional one. Personally I have had practical experience of both kinds, and although each has its advantages and disadvantages I am inclined to ommend the roller, or full-length, shaft, fitted with tension apparatus shown in Fig. 42, because the is always ready for immediate use, and is not put of its course, as it is in Fig. 41.

CHAPTER VI.

THE DIVISION-PLATE AND INDEX.

WITHOUT this apparatus the capabilities of a lathe would be greatly restricted, not only in the matter of ornamental turning, but in regard to its general usefulness. If, indeed, an eccentric or geometric chuck forms a part of the available apparatus, a good deal of handsome work can be effected, but this is owing to the fact that these chucks carry their own dividing apparatus. A division-plate upon the mandrel pulley, however, still further extends the capabilities of these and other compound chucks, and it is, therefore, always considered an essential of the lathe for ornamental turning.

In all first-class lathes the division-plate is of brass or gun-metal, and it is attached to the face of the pulley of which it practically forms a part. Circles of small holes are drilled in this plate, the numbers of holes being selected by their divisibility into equal parts ; but in some cases prime numbers are added to meet the necessities of clockwork. Of course the numbers must depend to a great extent upon the size of the division-plate, but in a lathe of five-inch centre there is room for five circles of holes, the outer containing 360, which is divisible by 2, 3, 5, 6, 8, 9, 10, and many multiples of these up to 180. A second circle will contain 192 ; a third, 144 ; a fourth, 120 ; a fifth, 112.

If, however, the pulley is smaller, as in a four or three-inch lathe, the outer circle will probably be 120 or 96, and inner circles will be 84, 60, 48, 12, with, perhaps, 7 as a prime number; but any number likely to be useful can be supplied by the maker.

The index may be merely a stiff, flat spring (Fig. 43), with a projecting conical point at its upper end, carefully turned to fit well into the holes in the plate, and having at its lower end a short cylindrical pin,

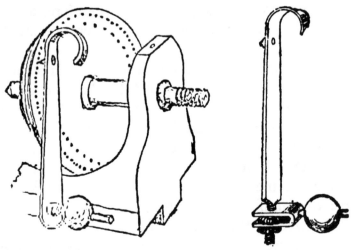

FIGS. 43 AND 44.

which enters the hole in a globular boss screwed into the base of the headstock.

This amply suffices for general purposes and for most ornamental work; but, in certain cases, it is advantageous to substitute an adjustable index (Fig.). This is so arranged that after the index point placed in any desired hole of the plate the spring be raised or lowered slightly by turning a milled The effect is to shift the pulley a minute degree in one or the other direction. The object of this is to adjust the point at which a pattern is to be commenced on the work when such point is not in exact accordance with one hole of the division-plate, but falls between two of the holes. It may happen that the use of one of the other circles of holes may bring matters right; but such circle may not be divisible by the same numbers as the first, and to shift from one circle of divisions to another after the work has begun is not always an easy matter. The extra cost of the adjustable index is very slight, so that it is generally expedient to have it fitted instead of a plain one.

The tedium of counting and setting the index after each cut of an extensive pattern is very great, and it is no wonder that even those devoted to this class of work get tired of it, and much ingenuity has been exercised in devising a mechanical count-plate. The difficulty has been satisfactorily overcome, but, unfortunately, the apparatus is too costly for adoption in simple ornamental turning such as we are describing, and it must remain one of the unattainable luxuries for those of moderate means.

It is in this respect that a five or six-inch lathe is so superior to one of three or four-inch centre. To peer closely into so small a division-plate, drilled into the face of the cast-iron pulley, is exceedingly trying to the sight; whereas the larger plate, which is generally of brass, is much more legible. I have sometimes fancied that a thick plate, with the edge notched into sloping teeth, like those of a clock 'scape-wheel, and a spring detent that would cause an audible click as it passed over each tooth, would form a good counter, as it is easier to count by ear than by sight ; but I have not personally tried the plan. The cog-wheel, however, attached to the pulley of a back-geared lathe, especially if the teeth are machine-cut, forms a very efficient division-plate if an index is added. This index is only a steel spring with a cross or T-piece at the top, filed to fit accurately between the teeth of the wheel. The flat of the spring is, of course, in this case parallel to the lathe-bed, otherwise it is fitted in the same way as that already described. Of course, in such case we are limited to the number of teeth ; but as the work to be ornamented is almost certain to be very much smaller than the toothed wheel, the latter will enable ornamentation of a sufficiently fine character to be done. In any case, if a back-geared lathe has to be used, there is sufficient width in the turned part of the *face* of this wheel to receive a goodly number of holes ; and it should be always drilled as a division-plate, whether or no the actual cogs are brought into use for that purpose.

CHAPTER VII.

THE ECCENTRIC CHUCK.

THE eccentric chuck is represented complete in Fig. 45, and although it is somewhat expensive—costing anything from £3 to £20, according to size and maker—it is so useful a chuck that I feel compelled to include it among the apparatus for simple ornamental turning.

This apparatus consists mainly of two parts—the foundation-plate, which screws on the nose of the mandrel; and the slide, which carries a similar screwed nose upon which the work is carried. For ornamental turning, however, there is an intermediate fitting forming the division-plate of the chuck, which fitting also carries the screwed nose. The eccentric chuck, used by itself, enables the centre of the work to be moved away from the general line of lathe centres, bringing a new spot into that line. This is effected by the main screw, which causes the slide with the division-plate and nose-piece to travel downwards. This power of placing work eccentrically is often of value, apart from ornamental turning. For the latter, except in a limited number of cases where the eccentric cutter fulfils the need in a different manner, it must be considered essential.

The construction of the eccentric chuck is as follows—and here again I may remark that any good mechanic should be able to construct one for his own use if he is unable or unwilling to incur the cost of

E

the purchased article. First of all there is the back-plate (A, Fig. 46). This, which is a plain casting, is mounted face downwards on a face-plate, or other suitable chuck, and turned over the flat part and projecting boss, B. The latter is then bored and screwed inside with a tap and chasing tool, and the end of the boss faced carefully and made quite true. It is then reversed and screwed on the mandrel, and the flat is turned and faced with great accuracy to a dead level. All this is simple and easy lathe-work that should present no difficulty to any one used to slide-rest tools, and who is a fair hand at metal turning. To make, however, a perfect job of this and similar work it should be further levelled by scraping—being tested on a surface-plate. This is a plate of cast iron which has itself been worked by planing and scraping to an absolutely dead level. It is smeared with red-lead or ochre made into a thin paste with oil, and a very thin coat is spread on the surface-plate. The part to be tested is laid upon it and moved with a circular motion, when any high places marked by the colouring matter. These are down patiently, one by one, until it is seen the colouring matter marks evenly the entire face, showing that no high spots remain. This operation is tedious, but is alone capable of producing perfectly true surface. In the present case, if turning has been very carefully done with a good slide-rest, and light cuts taken with a suitable oil, this scraping will not be absolutely necessary; as, indeed, an unskilful hand may do more harm than good.

Upon the face of this back-plate thus levelled two

bars of steel, C, are fixed, between which the front-plate has to slide. The bars are filed very carefully to a V-shaped or double-bevelled edge, and must be fixed perfectly parallel to each other by two or three countersunk screws. One bar—the farthest from the operator when the chuck is vertical—is made with the screw-holes slightly elliptical, so as to render it

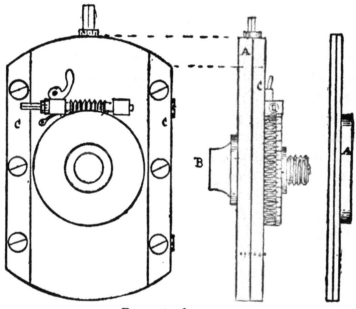

FIGS. 45, 46, AND 47.

capable of slight adjustment in the direction of the other bar, and two set screws, with heads overlapping the bar, are tapped into the edge of the back-plate, by means of which the sliding-plate may be made to move with greater or less stiffness between the bars. This sliding-plate is cast with a boss upon its face (A, Fig. 47). Such boss, when turned and faced, forms a bed for the dividing-plate, hollowed out underneath, to fit nicely over it : it need only stand up a quarter of an inch above the face of the plate.

This plate now has to be faced on the back and front, and both edges nicely grooved to slide between the bevelled edges of the guide-bars. It must, therefore, be filed, first of all, to the exact width required ; its two sides being made absolutely parallel. To turn the flat face, mount it in a wood chuck by the projecting boss, or fix it in an American self-centering chuck, gripping it by the boss. The utmost care must, however, be taken that it beds down and runs truly, which a light trial cut will determine. One great secret is to have a tool that really cuts—a round-end one is the best—and not to attempt deep cutting which would probably tear the work out of the chuck and ruin it at once. The plate not being circular increases the necessity for the utmost caution. When this lower face has been satisfactorily surfaced, the other, with its boss, must be attacked, and this boss has to be so turned as to be absolutely central to the plate in both directions. To ensure this it should be drilled with a small central hole before it is removed from the chuck in which it was held while the flat face was being turned. A similar hole, drilled in the back-plate already finished, will allow a steel pin to be driven through both, and will insure the true centering of all parts.

If now the second plate is in position between its bars, with a pin through the hole just spoken of, and a steady pin near the bottom (which is always fitted), the plate can be turned *in situ*, and the boss also finished, and all will necessarily be true. But before this can be done, the edges of the slide have to be

grooved to fit the chamfered bars. A fairly easy way
to accomplish this, if no planer or revolving cutter is
available, is to block the plate up on the tool-holder
of the slide-rest, laying the finished side downwards
upon a plate of metal or of wood, and then to drill
the groove by using the traverse. The drill must be
a flat one at the end, ground to such an angle as to
form the groove required ; and when this is finished,
a further cut should be made with a small drill in the
extreme angle. This will not only form a reservoir
for oil, but will allow the chamfered bars to bed closely
into the V grooves. A reverse process, which will
have the same result, is to chuck the work up on an
angle-plate, fixed to the face-plate of the lathe, and
fix a drilling instrument in the slide-rest. In this
case the division-plate and index is used to hold the
mandrel, and set the angle-plate exactly horizontal.

After drilling the V grooves in the edges of the
sliding-plate, they will need filing or scraping to a
more even surface than can be given by the drill
alone. In the present day this work may be beautifully done by one of those handy little shaping or
planing machines offered by one or two makers at
£5, or thereabout ; but it is as well not to suppose
the reader is the happy possessor of one of these.
They greatly simplify such work as is under consideration ; but, at the same time, excellent work was
done long before these luxuries were invented.

The angles of a 3-square saw file are each 60°, and
if the drill has been ground to this angle, the grooves
may be perfected by this well-known tool. The tang,
when heated, can be bent so as to set the file at an
angle to its handle, which will facilitate the operation,

and it is not a bad plan to break the file across so as to leave only about an inch to work with. For smaller touches, where the file would encroach too far, a small three-cornered scraper will also prove an efficient instrument. The final work consists of grinding with oilstone crushed to powder, and used with oil. The slide is placed between its guide-bars, and these are gradually set forward till the movement of the slide is smooth and equable.

The nose-piece forming the division-plate should present no great difficulty as regards turning. The casting, Fig. 50, is first held by the projection which will form the screw to carry the chucks, and faced up truly; and a recess is then to be turned to fit accurately upon the projection on the sliding-plate. It must then be drilled with a slightly conical hole, as it turns on a central pin fixed on the sliding-plate. This pin is of steel, and should be $\frac{1}{4}$ in. to $\frac{3}{8}$ in. thick, and must be fixed accurately, perpendicular to the plate. Its length depends on that of the chuck-nose, which is usually about an inch long. This chuck-nose is the part likely to present difficulty, as the screw *must* be cut in the lathe while the chuck is in position, or it will not be absolutely true. In addition, the count-wheel, or division-plate on the same casting, must either have its edge cut as a cog-wheel, or it must be formed as a worm-wheel, to be driven by a tangent screw. This is really work for a self-acting lathe, although it can be managed without by any one who is a prac- tised turner, and can command the necessary appa- ratus. In very many cases it would be better to and this bit of work to a professional, letting him it the teeth or worm, and also the nose for chucks, which can, if necessary, be subsequently skimmed

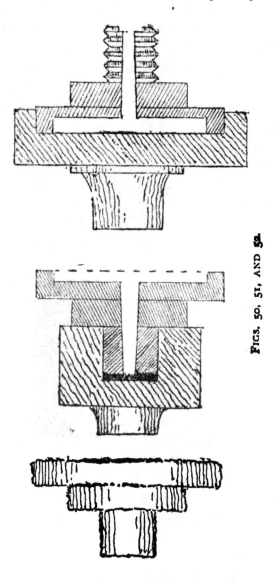

FIGS. 50, 51, AND 52.

over with a chaser when the chuck is otherwise complete and in its place upon the mandrel ; but probably

very slight correction in this respect will be needed to bring it into absolute truth. If, however, the reader feels too independent to apply for professional assistance, he had better go to work in the manner following:

Mount the casting, as stated, by the nose-piece and turn it over as described. Remove it from the chuck and remount on a wooden or self-centering grip-chuck by the part turned, so that the nose comes outward : turn this truly, leaving it only a little larger than the finished size ; drill for the centre-pin, if not already done ; recess the face of the nose-piece just deep enough to receive a thin washer and the head of a small screw, and drill and tap a hole to receive this screw, which is to retain this casting upon its central pin.

All is now ready for the nose-screw and for cutting the teeth of the division-plate, or racking its edge for a tangent screw. But for the latter purpose it is necessary to get at the *edge* of the dividing wheel which involves chucking the piece again by the nose, which is better done before the screw is cut upon it, or the latter may be damaged. This screw may, however, be cut first and a similar internal screw cut in a boxwood chuck to receive it, when, if all is done well, it should run quite true.

Still, I should prefer to hold the casting by the nose-piece before cutting a screw upon it, and let the wheel-teeth, or rack, have the first attention. To cut these an overhead is necessary, and a horizontal or universal cutter frame, with a tool of the form of interval between any two teeth.

Still better (if the instrument is so fitted) is a small wheel-cutter to mill out the teeth, as the fly-cutter will frequently stick in the cut unless the greatest care is taken to advance it slowly. If a *geared* cutter is in stock so much the better, as it has greater power ; but it is not a matter of necessity.

The cutter, fixed as usual in the slide-rest, must be carefully adjusted to the exact height of centres, and each cut made by slow traverse to and fro of the slide-rest ; it is not at all a difficult job, but must not

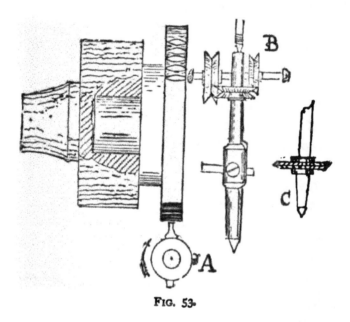

FIG. 53.

be unduly hurried. Of course the distance between the teeth is regulated by the division plate of the mandrel pulley.

Of the accompanying illustrations, Fig. 50 represents the untouched casting ; Fig. 51, the same in a wooden chuck, drilled and recessed to fit over the turned boss on the sliding-plate. It is then reversed and held in another chuck by the turned part, faced and the nose screwed to the same pitch as the mandrel, as Fig. 52. In Fig. 53 it is seen as again reversed and held by the nose—here represented without the screw. This brings the part that is to become the division-plate into position to be acted upon by the fly-cutter, A, and thereby to be cut into teeth for a click-wheel or for a worm-wheel ; B shows the revolving spindle of the horizontal cutter, with pulley and guide wheels; and C, a similar spindle, with wheel-cutting mill, which is a substitute for the fly-cutter.

A properly-shaped cutter will entirely complete the teeth of such a wheel ; but if a worm-wheel with tangent screw is preferred, the wheel is first cut with a fly-cutter or mill, and finished by being revolved in contact with a hob which is a parallel screw-tap, grooved lengthwise in several places to give it more of a cutting action. Both tap and wheel revolve, and the grooves take the exact form of the thread of the hob, which should be a duplicate as to size and pitch, of the tangent screw which is to be ultimately used as the driver. This is more of a punching than cutting operation, but it effects its purpose perfectly, and the advantage of a worm wheel and tangent screw over a click wheel is great. In the first place, this arrangement is self-locking, whereas the click or catch needs a spring to keep the detent in close contact with the wheel, and a sudden strain upon the latter, such as is produced a deep cut, will often cause it to rotate, damaging teeth, and frequently spoiling an otherwise well-pattern. There is also no backlash, owing to the contact of the wheel with the screw ; and the is so smooth and equable that there is little fear of a miscount—the use also of one hand only is necessary instead of two. The tangent screw is fitted in a brass casting (Fig. 54) pivoted on a screw and kept in contact with its worm-wheel by a cam. The frame is kept away from the wheel by a spring when the cam is so turned as to free it, in order to allow the wheel with the chuck and work to be moved several divisions at a time ; but when the ornamentation is in progress and only a slight rotation is needed after each cut, the tangent screw alone is used. Finer dividing can be done in this way than by using a click-

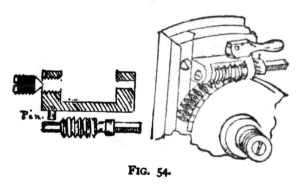

FIG. 54.

wheel, and the additional cost is not great, nor is it much more difficult to make.

It may as well be mentioned here that variously-shaped springs very suitable for this work can be obtained at an ironmonger's or gunsmith's, as many shapes of such are used in locks, guns, and rifles ; and spring-making is hardly within the scope of an amateur's capabilities.

The foregoing details will, I think, suffice to enable an amateur to make himself a sufficiently good chuck for simple eccentric turning, if he is already a tolerably good hand at metal-work. At the same time, I do not altogether advise the home-construction of lathe apparatus, which needs special accuracy and perfect fitting of each individual part. Such work is rather for a master-hand than for an apprentice, and the purchased article will save much disappointment in the execution of ornamental work.

In speaking in an earlier chapter of the eccentric cutter, which is very often used in combination with the eccentric chuck, I omitted to state one very useful and curious fact in connection with it, namely, its

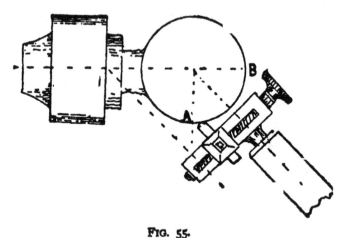

FIG. 55.

capability of cutting an accurate sphere in boxwood, ivory, or other suitable material. To cut a perfect sphere needs two operations, as the work has to be chucked twice, which will be evident when describing the formation of the hemisphere. The work is first turned by hand to the hemispherical form, and may, at the same time, be made to approach the perfect sphere, leaving, however, a fairly substantial neck, which will be subsequently cut away. The eccentric cutter is now fixed in the slide-rest, and its frame set to 45° with the lathe centres or bed of the lathe—and set exactly to bring a round-nosed tool to height of centres. The cord is to be removed from the flywheel and the lathe pulley slowly rotated by hand, while the cutter is rotated at full speed. By means of the slide-rest, the tool is gradually advanced so as not to hitch in the work, and the result will be a perfectly-formed hemisphere. The tool is set to touch the two points A and B (Fig. 55) in its revolution. Of course, if a sphere is required, the piece has to be re-chucked by the finished hemisphere, and the operation repeated on the second half of the work.

Having described the apparatus used for the more simple decoration of articles turned in the lathe, it only remains to add a few words in explanation of their several uses and the principles underlying what are known as eccentric patterns. These consist in every case of intersecting circles, or segments of circles, variously placed so as to form a pleasing whole. These are for the most part cut with double or single-angled tools placed in the eccentric cutter, but some are grooved out with a drill to which linear as well as rotary motion is given. In order to place these circles as may be desired, three several or combined means are at our disposal—the division-plate of the mandrel, the movement of the slide-rest, and the linear and circular movement of the eccentric chuck. With these we are enabled to bring all parts of the face of the work under the action of the cutting-tools.

Let Fig. 56 represent a block of hard wood—box, cocus, blackwood, and ebony, are mainly used for this work. The block must be very accurately faced with a round-nosed tool in the slide-rest, or the cuts made subsequently will be deeper in one part than in the other, which would be fatal to success. Fixing the mandrel by the index-peg in the zero-hole of any one of the circles of holes selected, place the eccentric cutter in the slide-rest, set to produce by its rotation

Fig. 56.

small **circle,** using a keen double-angled cutter. It of **course** presumed that the cord or cords from the verhead **are** suitably arranged to drive the cutter at good speed—all which must be done and tested efore a cut is **made.**

the movement **of the** slide-rest alone in a line

across the face of the work we can place a succession of intersecting circles in a straight line across a diameter, as seen in the Figure—or, as in Fig. 57, leaving the slide-rest in its first position and using the division-plate of the mandrel alone, we can place such circles all round the circumference, closer or

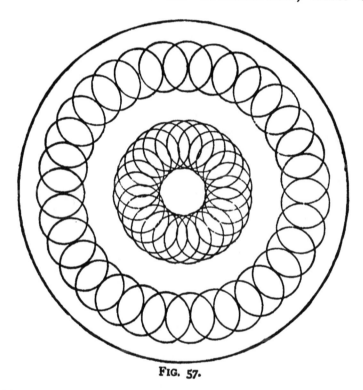

Fig. 57.

wider apart according to the number of holes passed over before each fresh cut. After one circle is cut we may shift the slide-rest, and thus bringing the tool more central we can cut a second, or any desired number of these interlacing circles. We may also place them radially (Fig. 58), which is but a repetition of Fig. 56, but with the mandrel shifted by its index a certain number of holes after each **row of** circles has been cut, which will bring into the **plane** of the line of centres a fresh diameter at each setting. But we cannot by the slide-rest and division-plate alone produce such a pattern as Fig. 59, because the two circles of interlaced design are each round **an**

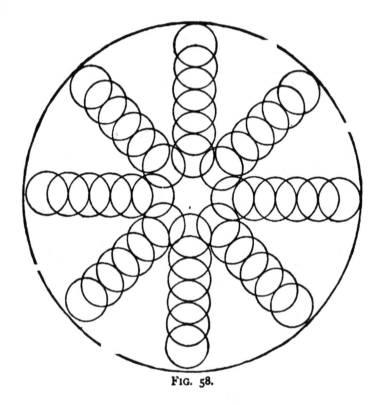

FIG. 58.

independent common centre, whereas in the previous designs the main centre of the disc is the common centre of the interlaced series of circles. It is here that the eccentric chuck comes into use enabling us to bring the new centres into the line of centres, and then, using the division-plate of the chuck to place the several interlacing circles, we can cut them as before with the eccentric cutter.

In the pattern known as the Turk's Cap, the circles have a diameter of half the radius of the main circle (Fig. 60), through the centre of which all the cut circles pass. This is commonly seen on the backs of

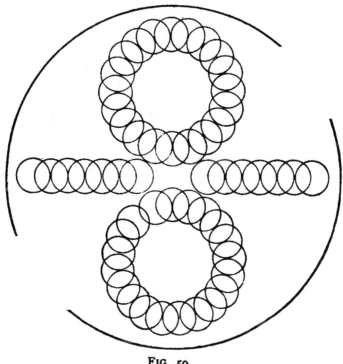

FIG. 59.

watches. For this only the eccentric cutter and division-plate of the mandrel come into use—the former regulating the size of the circles, the slide-rest their centres, the index-plate the spaces between them. The illustrations being drawn by hand, are of course only indicative of the *principle* of this kind of

F

work, and are consequently but very coarse, but they show with greater clearness the details of the several designs. The Turk's Cap and many other patterns may be cut—by means of the eccentric chuck alone—with a fixed tool in the slide-rest, in which case the overhead apparatus is not used, as the work itself

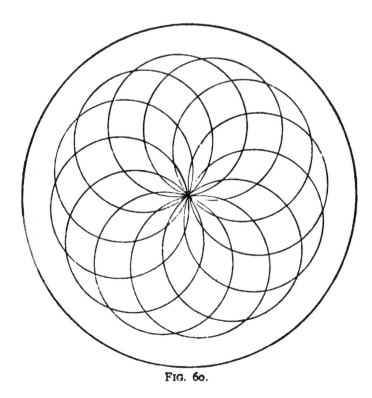

FIG. 60.

takes up the several positions necessary, the tool merely needing withdrawal after each cut.

In actual practice these designs do not deserve the opprobrium which is often laid to their charge of being nothing but scratches not worth the trouble of cutting, because such is far from being their real condition. If, indeed, they really are "scratched," the fault is the workman, who has not been at the trouble of sharpening and burnishing the tools. If these are in perfect order, each cleanly-cut surface will reflect the light like an infinite number of mirrors, and the effect will be exceedingly beautiful. They can never be adequately represented by printing, because of the absence of this play of light and shade. Moreover, a block which is intended for printing needs very shallow cutting, or the printed lines are so thick and clumsy that the beauty of the pattern is utterly lost. I have endeavoured to obtain some photographs, however, of work of this kind cut by me many years ago, before hand and eye began to fail, and I hope these may turn out sufficiently satisfactory to illustrate simple ornamental turning. Fig. 61 is added to show about how close together the lines forming a pattern are usually cut. The outside is called a shell pattern, and the whole is cut by the aid of an eccentric cutter and by the movement of the division-plate by which the shells are formed. These, if examined, will be seen to be merely a succession of circles, the centres of which are upon a line forming a common diameter to all of them; but each being smaller (or larger) than the one next to it causes all the circles to touch at one point. How is this done? Evidently by shifting the centres after each cut by means of the slide of the rest, while the eccentric cutter-slide is similarly shifted to enlarge or diminish the size of the circles. If again, as here, the shells are made in a circle round the pattern, they are placed in position by the division-plate on the mandrel. Each shell is not, as a rule, individually cut, and then the next begun, but all are cut simultaneously. A ring of circles is first cut, using the division-plate to place them either in contact, as here, or at any given distance apart; then the cutter-slide is shifted to reduce the size of the circle, and the slide of the rest is moved to shift its centre nearer to the common centre

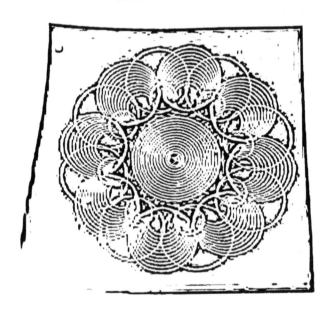

FIG. 61.

···· and, with the same settings of the ···· as were first used, a second ring of ···· and this is repeated till the pattern is ····. These shells should be cut with a *single-tool*, well polished. The Turk's Cap (Fig. 60) ly consists of a ring of large circles, the circ···ce of each passing through the common Their size is determined by the distance the cutter is set from the centre of the instrument ; their position by the slide-rest ; their distance from each other by the division-plate of the mandrel.

I think it will teach the reader the whole art of ornamental turning (*i.e.*, of surface patterns) to cut successively the examples here given. They may be cut as coarsely as the illustrations, and will thus save the operator the tedium of finer work. The object is solely to teach the learner which of the means at hand produces the several results—size of circles ; position of the same ; and the best sequence of the operations, namely, where to begin, and how most safely and expeditiously to follow on.

In the photograph (Fig. 62) accompanying this chapter the third of the flat discs shows fairly well the eccentric work on the original. The cutting is fine and even. I forget the exact number of circles which form the border, but probably there are over one hundred, and the rest of the pattern must have contributed to tax severely the patience of the writer, who, however, in those far-off days had more zeal than discretion.

The central ornament and ivory knick-knack upon it show chiefly fluted and ordinary drilled work. The flutes were cut after the ivory had been partly hollowed out, but were not cut through. The hollowing was then completed until the flutes were cut into, completing the design. This is often the safer mode of working and produces a better result. It also removes the danger of breakage by the too rapid penetration of the drill.

In the frontispiece (Fig. 63) are shown a few more examples of simple ornamental turning not even

FIG. 62.

needing an eccentric chuck. All the fluting is done with a plain round-ended drill, which tool almost without other appliances is capable of producing very beautiful work. The tall pedestal of the article on the right-hand side has a bowl with six flat sides. These are done with the eccentric cutter after the bowl is turned. It is left thicker than would otherwise be the case, and the number of sides are determined by the division-plate on the mandrel pulley. This flat-sided work is often needed, as it adds greatly to the beauty of a design by contrast with work of circular or elliptic outline. Each such flat surface may be subsequently decorated with drilled or fluted work, but a caution may here be given not to multiply decoration to too great an extent. The rule is " Decorate a construction, but do not construct a decoration." The twisted pedestal of the ornament on the left of the photograph was done entirely by hand ; the other, which is rather of a complicated design, was cut by a cord arrangement between the slide-rest and overhead, with a second overhead shaft driving a flying cutter. A self-acting screw-cutting lathe would have done away with the need of one overhead shaft, but such was not at the time in my possession. I must now bring to a close this somewhat imperfect sketch of the method and means of accomplishing simple decorative lathe-work. I have shown that up to a certain point it is by no means a costly amusement, and is capable of extensive development by a thoughtful and patient workman. Even if the eccentric chuck is omitted, which is the most expensive of the apparatus described here, a great deal of excellent and handsome work can be done. Like all good work in this world, Ornamental Turning needs energy and perseverance to carry it to anything approaching perfection, but it amply repays the care bestowed upon it ; and to business men especially, who have to toil drearily behind a desk from morn to dewy eve, the Lathe supplies that kind of healthy and intellectual amusement so needed to keep the *mens sana in corpore sano.*

PARTICULARS OF MR.
GUILBERT PITMAN'S
LATEST TECHNICAL
PUBLICATIONS WILL
BE FOUND OVERLEAF.

85, FLEE
LONDON,

BOTTONE'S

Amateur Electrician's Workshop.

Two Handbooks of Practical Instruction in the making of
Electrical Models and Appliances.

Book I.—Illustrated by 30 carefully-prepared diagrams.

CONTENTS.—How to Make an Electro-Motor—The Making-up of Lalande-Chaperon and Chromic Acid Batteries—The Construction of a One-inch Ruhmkorff Spark Coil—How to Make a Wireless Telegraph—How to Use a Wireless Telegraph—Sparking Coils for Motors: How to Make and Repair.

Book II.—With 40 Illustrations.

CONTENTS.—How to make a Wireless Telephone—How to Make a Wimshurst Electrical Machine—How to Build an Electric Despatch Boat—How to Make a Four-Volt Accumulator.

WHAT THE PRESS SAYS:

" Well worthy of a place in the library of the amateur."
Westminster Gazette.

" Gives the minutest directions as to the way to set about making an Electric Motor, Wireless Telegraph, etc., etc."—*Daily Graphic.*

" Capital ; . . . quite up to the author's high standard."
Literary World.

" A wonderful amount of practical information."—*Electricity.*

" Will be found most useful."—*Daily News.*

" Valuable advice."—*Liverpool Mercury.*

" Details for making a Wireless Telegraph are clearly set forth. Mr. Bottone is an 'old hand' at explaining the mysteries of model-making. . . . Amateurs will welcome this edition to his books."
Ironmonger.

" Mr. Bottone describes . . . in detail, in his lucid and comprehensive fashion."—*English Mechanic.*

" The instructions have the commanding merit of practicability."
Mechanical World.

" A practical little illustrated book."—*The Times.*

" Practical, smart, well illustrated, and clearly written."—*Scotsman.*

" Sure to be welcomed."—*Morning Post.*

" Noteworthy for the simplicity and clearness of its guidance."
Standard.

" Illustrated and practical."—*St. James's Gazette.*

" Should be a big demand for it."—*Birmingham Gazette.*

" The instructions are such that any novice could, with reasonable care, make the various items of apparatus described."—*Model Engineer.*

Each Book complete in Itself.

Price, bound in Cloth, 1s. 6d.; post free 1s. 8d.

Practical Kites

& Aeroplanes :

HOW TO MAKE
. . . AND . . .
WORK THEM.

Illustrated by Forty-three Carefully Prepared Diagrams.

Cloth, 1/6 net ; post free, 1/8.

By FREDERICK WALKER, C.E.,

Fellow of the Society of Patent Agents, Associate Member of the Aeronautic Institute.

A Practical Work on the Construction and Use of Kites and Aeroplanes, with special consideration of the best mode of constructing kites for ascertaining the state of the upper atmosphere, photographing, wireless and luminous signalling, and, for those possessing sufficient confidence, aerial flight or aerial traction.

"Should prove of immense interest. . . . Fascinated me vastly."
To-Day.

"Extremely useful manual."—*English Mechanic.*

12th THOUSAND.

Systematic Memory ;

OR, HOW TO MAKE A BAD MEMORY GOOD,

AND A GOOD MEMORY BETTER. . . . . .

By T. MACLAREN.

ENLARGED AND IMPROVED EDITION.

Complaints are continually heard about bad memories. Some have the misfortune of having been born with "shocking bad" memories; while others have, in their early years, been gifted with wonderfully good memories, but, by some mysterious process, have gradually lost the power of retaining even the most recent facts. Some, again, have the rare faculty of at once, and without the slightest difficulty, committing to memory everything they please, but, unfortunately, in a very few days all is entirely forgotten : others, on the contrary, have very great difficulty in fixing anything in their minds ; but when once a thing is fixed, it is fixed for ever. The truth is, that most persons do not know how to employ their memories. The system set forth in this remarkable work is intended to make bad memories good, and good memories better.

Price 1/- net ; post free, 1/1½.

Talking Machines and Records. . &

How Made and Used.

By S. R. BOTTONE.

**96 Pages. Crown 8vo. Fully Illustrated.
Price, strongly bound in Cloth, 1/6 net.
. . . Post free, 1/8. . . .**

The information conveyed in this book, by word and picture, will enable anyone to form an intelligent conception of the principles of the talking machine, and to make a record of any sound—be it noise or music, speech or howl— and to reproduce the same at the will of the operator.

Full and clear directions, accompanied by care-fully-prepared drawings, are given for making a simple but efficient and complete instrument. The historical portion is also fully dealt with, and the original instruments of Kratzenstein, Kempelen, Willis, and Scott, illustrated and described.

" Mr. S. R. Bottone's handbook on ' Talking Machines and Records,' just published by Guilbert Pitman, is an ex-cellent *résumé* of the *rationale* of the phonograph and gramophone, with full instructions for making a simple phonograph and the production of records and blanks. Everything is explained in the lucid and practical manner which always characterizes our old correspondent's com-munications."—*English Mechanic.*

" The important part which Electricity plays in the industrial progress of the world is now fully recognised—the future of Electricity is absolutely beyond the Realms of Imagination."

Professor J. A. FLEMING.

BOTTONE'S

ELECTRICAL ENGINEERING FOR STUDENTS.

Specially written in crisp, attractive, pictorial language, so that anyone can understand it, with full constructional details of all the appliances used, in order to meet the requirements of Home Students in Electrical Engineering, and also to fulfil the conditions of the City and Guilds of London Institute's Examinations.

By S. R. BOTTONE.

Author of "Amateur Electrician's Workshop," "Talking Machines and Records," "Ignition Devices for Petrol Motors," etc., etc.

This up-to-date, accurate, and comprehensive work is the most important book on Electrical Engineering ever issued at the price.

There are many students who, desirous of acquiring a practical knowledge of electrical work, find themselves hampered by their inability to see and to make the instruments of which they read in text-books. Many are precluded by circumstances of locality from availing themselves of the advantages presented by the Polytechnic Institutes; and some cannot attend these owing to the hours of attendance being inconvenient or impossible. This work is intended for the benefit of these, as well as for the general reader. In every case each step made in theoretical knowledge is illustrated by simple experiments performed with home-made apparatus, the construction of which, from specially prepared working-drawings, forms a unique feature of the book, and will enable the enthusiastic student to make and to understand the most useful pieces of apparatus employed by electrical engineers as well as if he had workshop training and experimental instruction.

PART I.—Magnetism and Magnetic Apparatus.

PART II.—Static Electrical Instruments, including a detailed illustrated description of the construction of a Wimshurst Electrical Machine specially for X-Ray work.

Crown 8vo. Strongly bound in Cloth. 49 illustrations. 160 pages. Price 2/- net.

AMATEUR'S COMPANION TO THE WORKSHOP.

Being a collection of practical Articles and Suggestions for the use of Amateur Workers, including detailed Instructions for the Home Construction of Electrical and other Models and Apparatus, forming entertaining and useful recreations for spare hours.

"The bow cannot possibly stand always bent,
Nor can human nature subsist without recreation."
CERVANTES.

By

S. R. BOTTONE,   J. H. WOOLFITT,
R. A. R. BENNETT, M A,   F. A. DRAPER,
THE EDITOR OF "THE ENGINEERING WORLD,"
AND OTHERS.

Amateurs will find this book instrumental in furnishing fresh and useful suggestions to be carried out in the workshop, and more particularly in providing suitable models and apparatus to be made with special reference to their ability to yield satisfactory results in the hands of the junior electrician; and to this end the minutest details are given as to the way to set about their construction, which in no case is beyond the powers of the ordinary amateur.

Copiously Illustrated by carefully prepared Working Drawings. . .

AMONG THE VARIOUS ARTICLES ARE :—
SMALL ELECTRO-MOTORS.
A CAMERA OBSCURA.
AN AMERICAN AERIAL YACHT.
AN ELECTRIC-SHOCKING MACHINE.
MODEL TELEGRAPHIC INSTRUMENTS.
STORAGE BATTERIES.
ELECTROTYPING.
AN ELECTRIC-BELL INDICATOR.
A SPARK-COIL AND CONDENSER.
TWO KINDS OF ELECTRICAL MACHINES. *Etc., etc.*

Cr. 8vo, strongly bound in cloth, price **1/6** net ; post free, **1/8**.

APPOINTMENTS.

"THE SUNDAY STRAND" says:—From the documentary evidence we have perused we are able to en.orse the words of "THE QUEEN" that

"The stirring promise of KENSINGTON COLLEGE that all qualified Students are ultimately placed in positions seems to be carried out with unfailing regularity."

KENSINGTON COLLEGE,
LONDON, W.

DIRECTORS.

MR. JAMES MUNFORD, *Member of the London Chamber of Commerce and the Society of Arts.*

MR. GUILBERT PITMAN, *nephew of the late Sir Isaac Pitman.*

Lady Superintendents: MRS and the MISSES MUNFORD.
Assisted by a staff of Lady Experts.

———✳———

A leading feature of this College is the preparation of Students in all subjects necessary for superior positions as Private Secretaries, Correspondents, Accountants, Book keepers, etc.

As soon as qualified, a remunerative appointment is offered each Student, the commencing salaries varying from about £65 to £104 per annum, and rising to £120 or £200 and upwards.

The Course comprises the following subjects, viz. :—

SHORTHAND,   TYPEWRITING,
BOOK - KEEPING BY DOUBLE ENTRY,
SECRETARIAL CORRESPONDENCE, and
TRAINING and ACTUAL EXPERIENCE of SECRETARIAL WORK
in the MODEL OFFICE ATTACHED to this COLLEGE.

MR. GUILBERT PITMAN (85, Fleet Street, E.C.) will be happy to send further particulars.

Write to-day for Prospectus.

N.T.